WEATHER R
LAB NOTEBOOK

DEVELOPED AT LAWRENCE HALL OF SCIENCE, UNIVERSITY OF CALIFORNIA AT BERKELEY
PUBLISHED AND DISTRIBUTED BY DELTA EDUCATION

FOSS Middle School Project Staff and Associates

FOSS Middle School Curriculum Development Team
Linda De Lucchi, Larry Malone, Co-directors; Dr. Lawrence F. Lowery, Principal Investigator
Susan Kaschner Jagoda, Denise Soderlund, Dr. Jan Woerner, Dr. Susan Brady,
Teri Dannenberg, Dr. Terry Shaw, Curriculum Developers
Dr. Kathy Long, Assessment Coordinator
Carol Sevilla, Graphic Artist; Rose Craig, Artist
Alev Burton, Administrative Support; Mark Warren, Equipment Manager

ScienceVIEW Multimedia Design Team
Dr. Marco Molinaro, Director
Leigh Anne McConnaughey, Producer and Principal Illustrator; Rebecca Shapley, Revisions Producer
Guillaume Brasseur, Producer and System Administrator
Dan Bluestein, Lead Programmer and System Administrator; Roger Vang, Programmer
Jerrold Connors, Senior Illustrator; Bonnie Borucki, Sue Whitmore, Illustrators
Alicia Nieves, Quality Assurance; Coe Leta Finke, Usability Review

Special Contributors
Marshall Montgomery, Materials Design; John Quick, Video Production; Ronald Holle, Content Consultant
John Jensenius, Warning Coordination Meteorologist, National Weather Service, Gray, MA
Warren Blier, Science and Operations Officer, National Weather Service Forecast Office, Monterey, CA
Patty A. Watts, Eric A. Pani, Water-Cycle Game
Learning about the hydrologic cycle and global climate change: A demonstration. Preprint
Fifth International Conference on School and Popular Meteorological and
Oceanographic Education: Weather, Ocean, Climate
Australian Meteorological and Oceanographic Society, Melbourne, 5–9 July 1999, 206–209

Additional Credits
Atmospheric Sciences, University of Illinois at Urbana-Champaign, WW2010 Project
Environmental Protection Agency; National Aeronautics and Space Administration
National Oceanographic and Atmospheric Administration; NOAA Climate Prediction Center
National Park Service, Johnstown Flood National Memorial
National Severe Storms Laboratory; National Weather Service

Delta Education FOSS Middle School Team
Bonnie Piotrowski, FOSS Managing Editor; Mathew Bacon, Grant Gardner, Tom Guetling, Joann Hoy,
Dana Koch, Cathrine Monson, John Prescott, Rebecca Waites

National Trial Teachers
Karen Burton, Cooper Middle School, Fresno, CA; Linda Stewart, Ahwahnee Middle School, Fresno, CA
Ted Stoeckley, Hall Middle School, Larkspur, CA
Laurie Erskine-Farley and Peter Josefsson, Parsons Middle School, Redding, CA
Lisa Evans, Southern Oaks Middle School, Port St. Lucie, FL
Gayle Dunlap and Donna Moran, Walter Bergen Middle School, Bloomingdale, NJ
Joan Caroselli, J. E. Soehl Middle School, Linden, NJ; John Kuzma, McManus Middle School, Linden, NJ
Terry Shaw and Joe Green, Irving Middle School, Norman, OK
Venus Ludovici, Janet McKenna, and Cheryle Jackson, Jay Cooke Middle School, Philadelphia, PA
Melissa Gibbons and Doris Taylor, Dunbar Middle School, Fort Worth, TX
Bebe Manning, Redwater ISD, Redwater, TX; Belinda Simpson, Redwater Middle School, Redwater, TX

LHS*

FOSS for Middle School Project
Lawrence Hall of Science, University of California
Berkeley, CA 94720   510-642-8941

Delta Education
...because children learn by doing.®

Delta Education
P.O. Box 3000 80 Northwest Blvd.
Nashua, NH 03063 1-800-258-1302

The FOSS Middle School Program was developed in part with the support of the National Science Foundation Grant ESI 9553600. However, any opinions, findings, conclusions, statements, and recommendations expressed herein are those of the authors and do not necessarily reflect the views of the NSF.

Copyright © 2003 by The Regents of the University of California
All rights reserved. Any part of this work (other than duplication masters) may not be reproduced or transmitted in any form or by any means, electronic or mechanical, including photocopying and recording, or by an information storage or retrieval system without permission of the University of California. For permission, write to: Lawrence Hall of Science, University of California, Berkeley, CA 94720.

Weather and Water   120-6523
2 3 4 5 6 7 8 9 BAB 07 06 05 04 03   1-58356-433-0

WEATHER and WATER
LAB NOTEBOOK
Table of Contents

Investigation 1: What Is Weather?
Class Weather Chart 1

Investigation 2: Where's the Air?
Air Investigations 3
Earth's-Atmosphere Questions 5

Investigation 3: Seasons and Sun
Sunrise/Sunset Times for 2000 7
Seasonal Changes 9
Response Sheet—Seasons and Sun 11
Beam Spreading 13

Investigation 4: Heat Transfer
Earth-Material Temperatures Chart 14
Earth-Material Temperatures Graph 15
Heat Conduction 17
Conduction through Materials 19

Investigation 5: Convection
Liquid Layers 21
Calculating Density 23
Response Sheet—Convection 25
Layering Hot and Cold Water 27
Convection Chamber 29

Investigation 6: Water in the Air
Relative Humidity 31
Response Sheet—Water in the Air 33
Dew-Point Questions 35
Pressure/Temperature Demonstration 37
Weather-Balloon Simulation 39
Upper-Air Sounding Graph 40
Temperature Number Line 41

Investigation 7: The Water Planet
Water-Cycle Game 43

Investigation 8: Air Pressure and Wind
Pressure in a Jar 45
Response Sheet—Air Pressure and Wind 47
Local Winds 48
Making an Anemometer 50
Pressure Map of the U.S. 53

Investigation 9: Weather and Climate
Solar-Balloon Observations 55
Reading Weather Maps 57
Response Sheet—Weather and Climate 59
Assessment—General Rubric 60

Name _____
Period _____ Date _____

CLASS WEATHER CHART

Date/time	Temp. (°C)	Air pressure (mb)	Relative humidity	Wind speed	Wind direction	Visibility	Other observations

Date/time	Temp. (°C)	Air pressure (mb)	Relative humidity	Wind speed	Wind direction	Visibility	Other observations

FOSS Weather and Water Course
© The Regents of the University of California
Can be duplicated for classroom or workshop use.

Investigation 1: What Is Weather?
Student Sheet

Name _____

Period _____ Date _____

AIR INVESTIGATIONS

Part 1: Record observations and questions.
While exploring air with a syringe, write three observations and three questions.

Observations

_____

_____

_____

_____

_____

Questions

_____

_____

_____

_____

_____

Part 2: Conduct an air investigation.

1. What do you want to find out? _____

_____

2. What materials will you use? _____

_____

3. How will you do this? Describe and draw. (Use the facing page to draw your setup.)

_____

_____

4. What did you observe? (Use the facing page.)

5. What did you find out? (Use the facing page.)

FOSS Weather and Water Course
© The Regents of the University of California
Can be duplicated for classroom or workshop use.

Investigation 2: Where's the Air?
Student Sheet

Name _____
Period _____ Date _____

EARTH'S-ATMOSPHERE QUESTIONS

1. What is the atmosphere? _____
   _____
   _____

2. Describe how the amount of air changes as you travel up through Earth's atmosphere.
   _____
   _____

3. Describe how the composition of gases changes as you travel up through Earth's atmosphere. _____
   _____
   _____
   _____

4. Describe how the temperature changes as you travel up through the atmosphere.
   _____
   _____
   _____
   _____

5. What layer of the atmosphere do you think is of greatest interest to meteorologists? Why do you think so? _____
   _____
   _____

6. What gases are found in the atmosphere? What gases are found only in the troposphere? _____
   _____
   _____

FOSS Weather and Water Course
© The Regents of the University of California
Can be duplicated for classroom or workshop use.

Investigation 2: Where's the Air?
Student Sheet

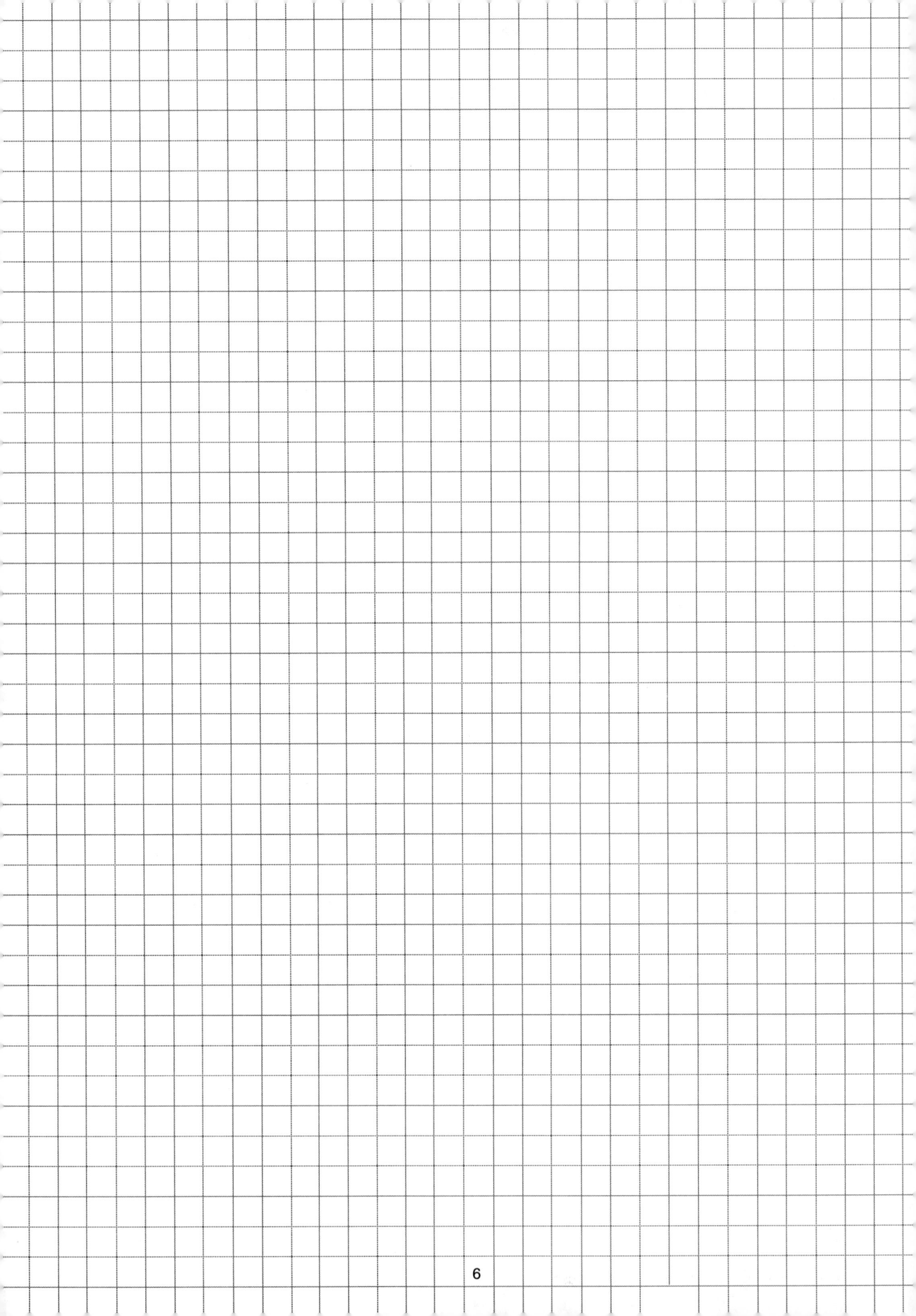

Name _____

Period _____ Date _____

SUNRISE/SUNSET TIMES FOR 2000

Sunrise and sunset times for the year 2000 in Berkeley			
Date	**Sunrise (a.m.)**	**Sunset (p.m.)**	**Hours of daylight**
January 21	7:21	5:20	
February 21	6:52	5:54	
March 21	6:10	6:22	
April 21	5:25	6:51	
May 21	4:54	7:17	
June 21	4:47	7:34	
July 21	5:04	7:26	
August 21	5:30	6:53	
September 21	5:56	6:07	
October 21	6:24	5:23	
November 21	6:56	4:54	
December 21	7:21	4:54	

Directions: Calculate the hours of daylight for each day and graph the results.

FOSS Weather and Water Course
© The Regents of the University of California
Can be duplicated for classroom or workshop use.

Investigation 3: Seasons and Sun
Student Sheet

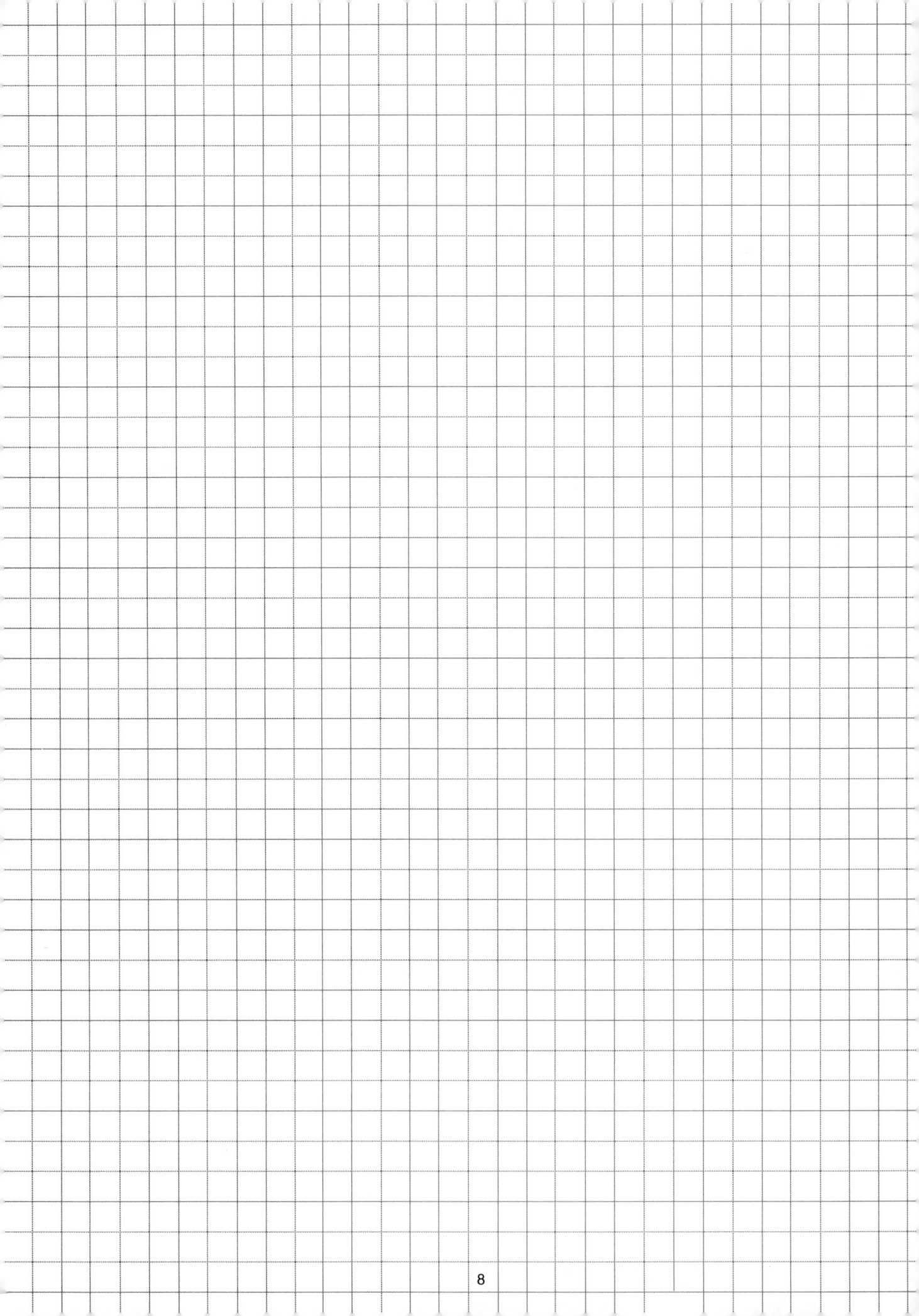

Name _____

Period _____ Date _____

SEASONAL CHANGES

Directions: Open the Seasons simulation. Select Berkeley from the first list of cities. Click month by month through the year, stopping at the equinoxes and solstices.

1. Circle the description that best describes the amount of light and dark experienced in a day by people living in Berkeley at the times listed below.

Spring equinox (Mar. 21)	More light	Equal	More darkness
Summer solstice (June 21)	More light	Equal	More darkness
Fall equinox (Sept. 21)	More light	Equal	More darkness
Winter solstice (Dec. 21)	More light	Equal	More darkness

2. Set the Earth View to "Side." You are now out in space looking at the Sun-Earth system (the same view seen in the Orbit View window). What does Earth look like at each of the times above? Draw a little picture of Earth in the boxes to the right. Show the parts of Earth in the light and in the dark. (Spring equinox is drawn already.)

3. What shape is the path traced by Berkeley as Earth completes one rotation? Describe and draw the shape. _____
   (To make sure, click the Advanced button and look at the North Pole in the Earth View window.)

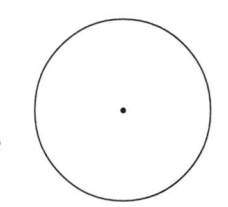

4. Where does the day/night line cross the Berkeley path at the summer solstice? Draw the top view, showing light and dark.

   How long is the daylight? _____

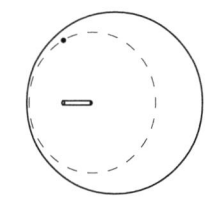

5. Where does the day/night line cross the Berkeley path at the spring equinox? Draw the top view, showing light and dark.

   How long is the daylight? _____

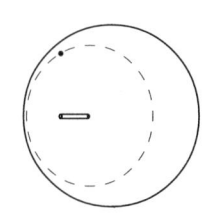

6. Select the cities between the equator and the North Pole one by one. Use the Earth View window to step through a day, hour by hour, on the summer solstice. Count and record the hours of daylight.
   Repeat the process for the winter solstice. Record the length of the day for each city. Graph the results. What is the relationship between latitude and day length?

FOSS Weather and Water Course
© The Regents of the University of California
Can be duplicated for classroom or workshop use.

Investigation 3: Seasons and Sun
Student Sheet

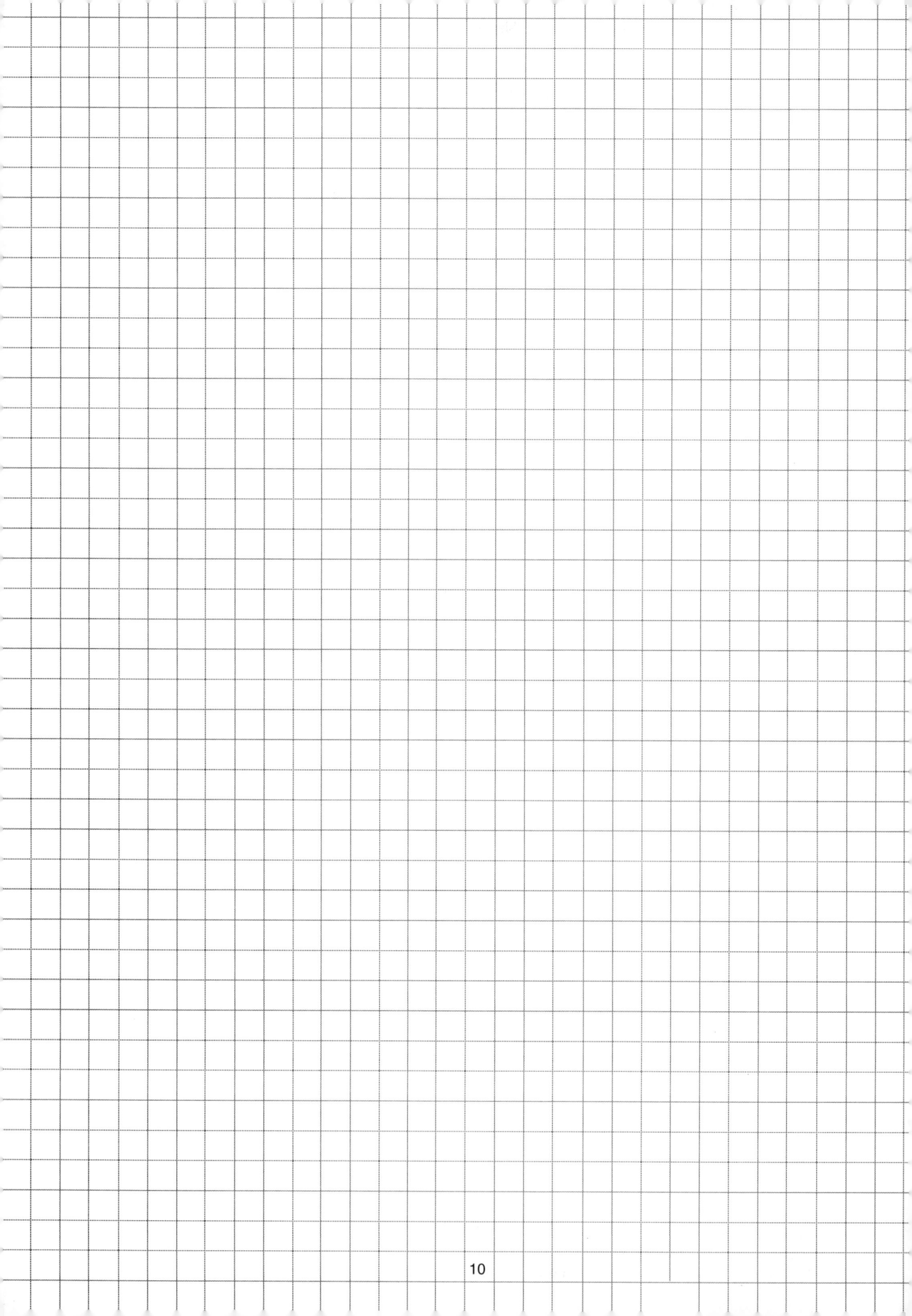

Name _____

Period _____ Date _____

RESPONSE SHEET—SEASONS AND SUN

Directions: Below are the journal entries of three students writing about the reasons for the seasons. Read each entry, then write a short paragraph explaining to each student what they need to change about their thinking.

Student 1 wrote: The reason for the seasons is that Earth revolves around the Sun in an elliptical orbit. When Earth is farthest from the Sun, it is winter. When Earth is closest to the Sun, it is summer.

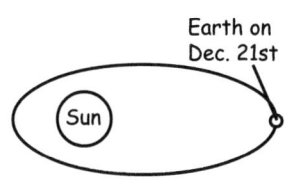

_____
_____
_____
_____
_____

Student 2 wrote: Summer is when we are facing the Sun, and winter is when we are facing away from the Sun.

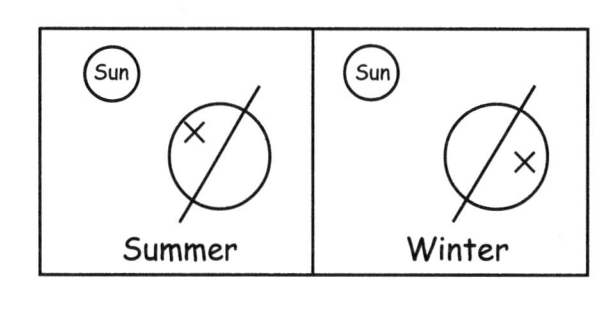

_____
_____
_____
_____
_____

Student 3 wrote: The tilt always leans toward the Sun. It takes 365 days for Earth to rotate one time. So when we are on the side toward the Sun, it is summer. When we are on the side away from the Sun, it is winter.

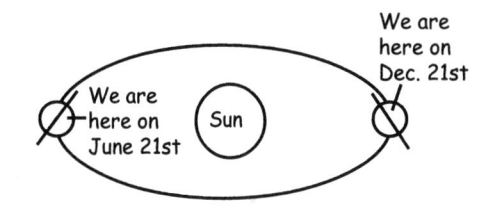

_____
_____
_____
_____

FOSS Weather and Water Course
© The Regents of the University of California
Can be duplicated for classroom or workshop use.

Investigation 3: Seasons and Sun
Student Sheet

Name _____

Period _____ Date _____

BEAM SPREADING

1. How do you explain the different shapes of the light spots?

   _____
   _____
   _____

2. When is the area of the spot largest?

   _____
   _____
   _____

3. Which spot delivers the greatest amount of energy to the floor?

   _____
   _____
   _____

4. If you put a penny in each light spot, explain which one will receive the most energy.

   _____
   _____
   _____
   _____

5. What influence does solar angle have on the heating of Earth?

   _____
   _____
   _____
   _____
   _____

FOSS Weather and Water Course
© The Regents of the University of California
Can be duplicated for classroom or workshop use.

Investigation 3: Seasons and Sun
Student Sheet

Name _____

Period _____ Date _____

EARTH-MATERIAL TEMPERATURES CHART

Air		Water		Soil		Sand		Time
Temp. change	Temp.	Temp. change	Temp.	Temp. change	Temp.	Temp. change	Temp.	3-minute intervals

FOSS Weather and Water Course
© The Regents of the University of California
Can be duplicated for classroom or workshop use.

Investigation 4: Heat Transfer
Student Sheet

EARTH-MATERIAL TEMPERATURES GRAPH

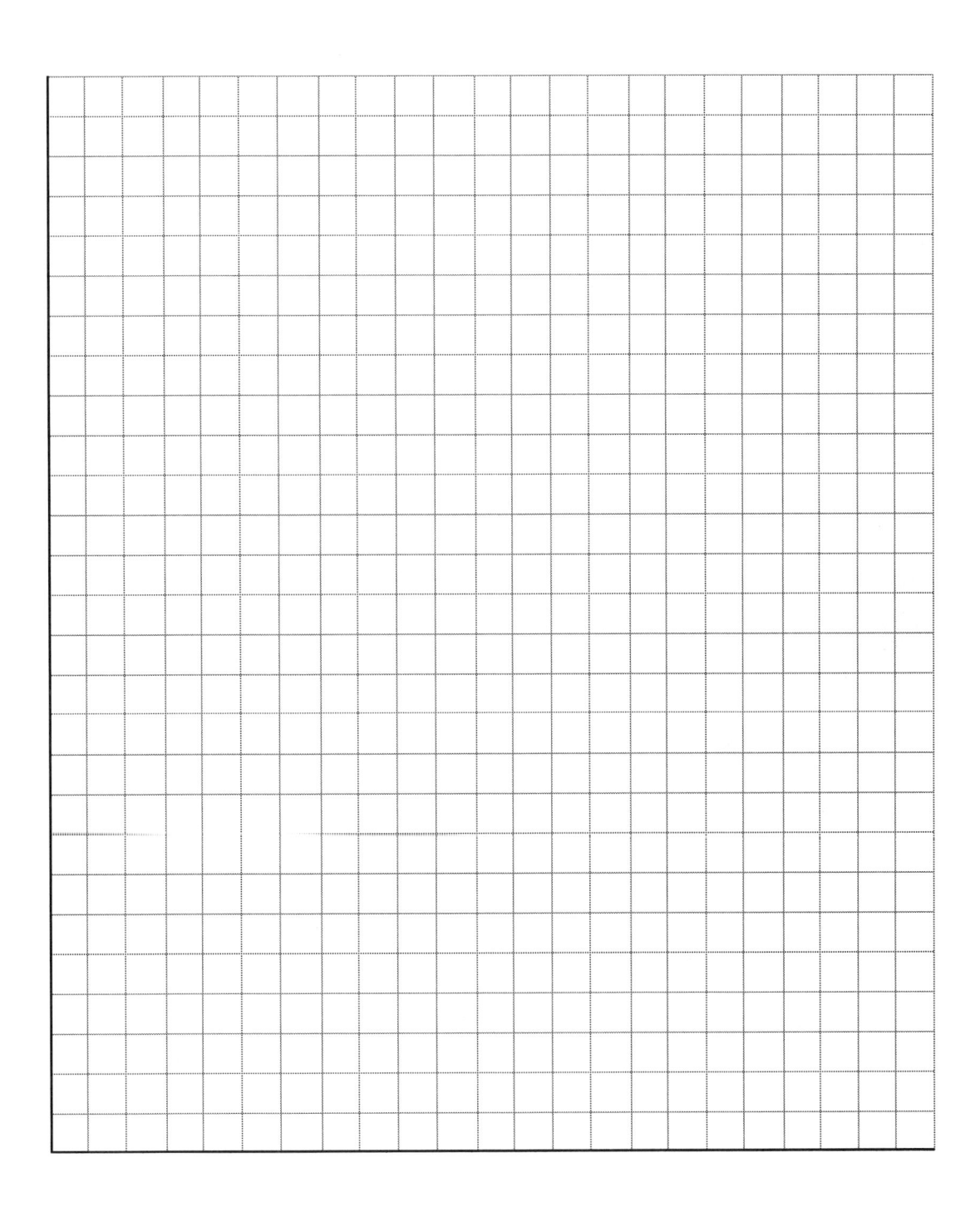

Name _____

Period _____ Date _____

HEAT CONDUCTION

1. Write a definition for heat.

   _____
   _____
   _____

2. Describe heat conduction.

   _____
   _____
   _____

3. Explain your understanding of how heat transfers from one material to another.

   _____
   _____
   _____
   _____
   _____
   _____

4. Explain why a soda can feels cold when you take it out of the refrigerator.

   _____
   _____
   _____

Name _____

Period _____ Date _____

CONDUCTION THROUGH MATERIALS

Materials
1 Steel bar
2 Temperature strips
1 Large clear cup
• Hot water
1 Aluminum bar
• Tape
1 Plastic-foam cup
1 Thermometer

Preparation and setup
1. Position a temperature strip on the steel bar with one end close to the end of the steel bar. Make sure the shiny side is up. Tape it in place.
2. Prepare the aluminum bar in the same way.
3. Fill the plastic-foam cup half full with hot water. Place it in a large clear cup for stability.
4. Place the bars in the water with the temperature strips up.

Observations and conclusions
1. Starting temperatures

    water _____   steel _____   aluminum _____

2. What happened when the metal bars with temperature strips were placed in the hot water? _____
   _____
   _____

3. Feel the two metal bars. How did heat get from the hot water to the temperature strip far above the water level? _____
   _____
   _____

4. Did the metals conduct heat? Which metal is a better conductor? Why do you think so?
   _____
   _____

FOSS Weather and Water Course
© The Regents of the University of California
Can be duplicated for classroom or workshop use.

Investigation 4: Heat Transfer
Student Sheet

Name _____

Period _____ Date _____

LIQUID LAYERS

Part 1: Layer salt solutions.
Find the sequence of colored salt solutions that will form clear layers.

1. Using the pipette, put a few drops of a colored solution into the straw. Try to layer a second color on top of the first color.

2. Use colored pencils to keep track of your results in the straws below. Circle the color combinations that produce layers.

3. Use the information to predict the order that will produce four colored layers.

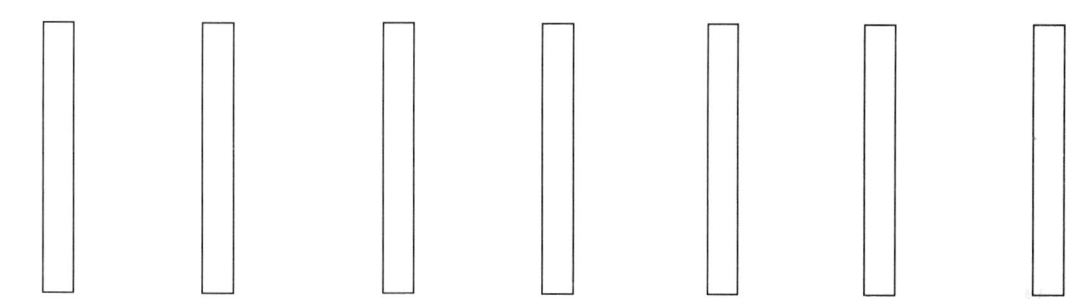

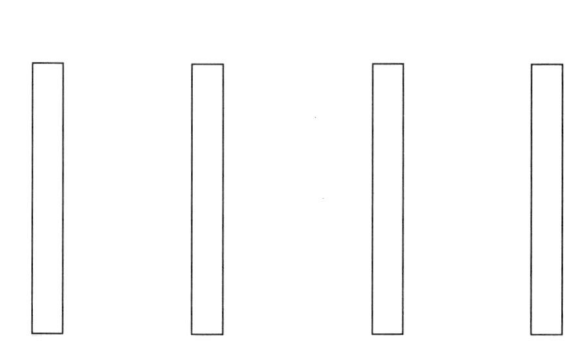

Color	Mass	Volume

Part 2: Explain salt-solution layering.

What do you think caused the salt solutions to layer in this way?

_____

_____

_____

_____

_____

FOSS Weather and Water Course
© The Regents of the University of California
Can be duplicated for classroom or workshop use.

Investigation 5: Convection
Student Sheet

Name _____

Period _____ Date _____

CALCULATING DENSITY

Write the equation for calculating density here.

Transfer the mass and volume data from the board into the table below. Calculate the density of the four solutions. Show your math work. Then solve the density problem at the bottom of the page.

Solution	Volume	Mass	Density
Blue			
Green			
Yellow			
Red			

Bianca and Joel mixed up a new salt solution and put in blue and red food coloring to make it purple. They then weighed 35 ml of the solution and found its mass to be 41 g. Where would the purple layer form if it were used with the four solutions above? Show your math.

The density of the purple solution is _____

The purple solution would form a layer _____

_____

FOSS Weather and Water Course
© The Regents of the University of California
Can be duplicated for classroom or workshop use.

Investigation 5: Convection
Student Sheet

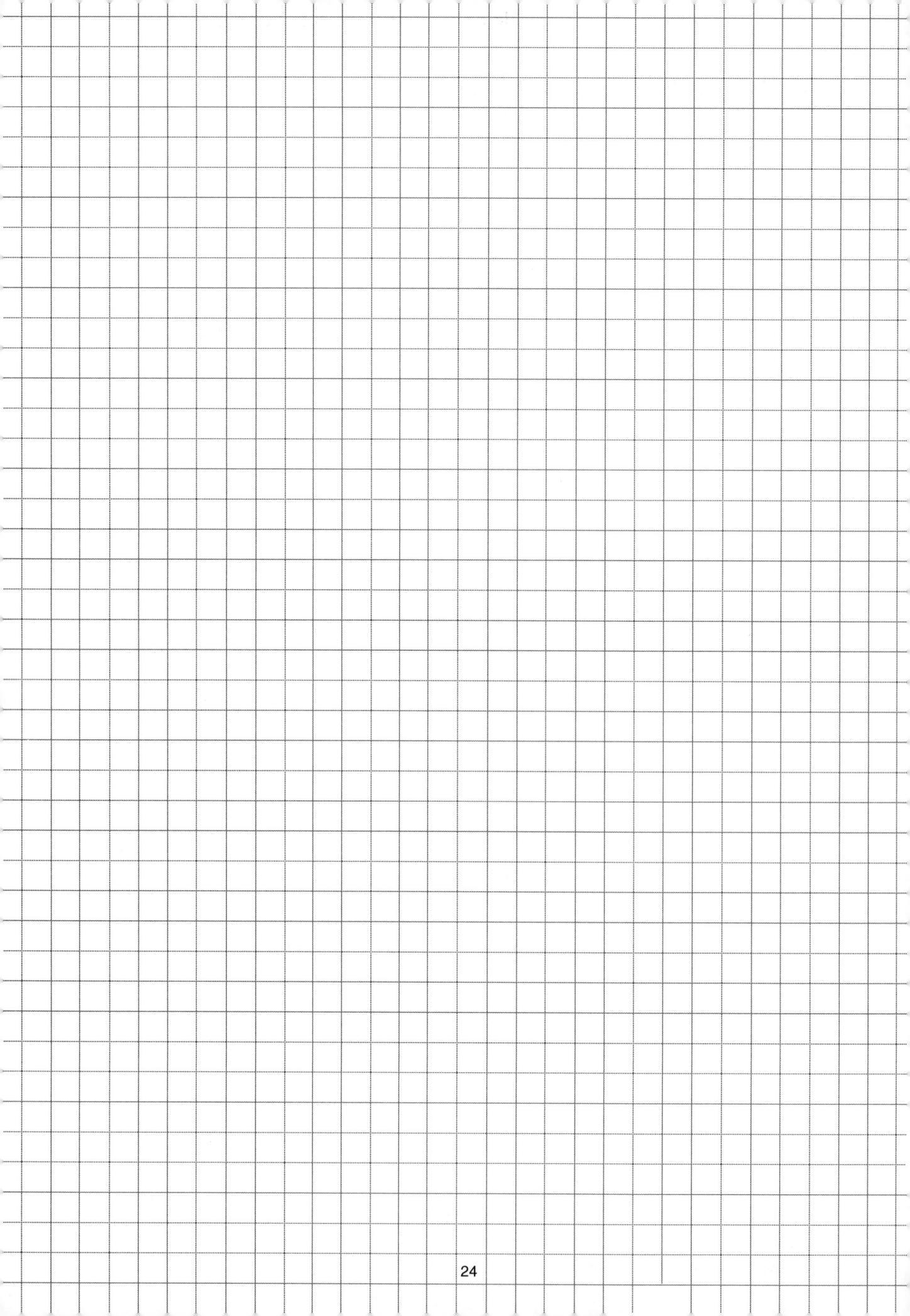

RESPONSE SHEET—CONVECTION

Rico wanted to make a shake-up toy for his little sister.

He had some little plastic stars and leaves. His plan was to put the stars and leaves in a jar and fill it with liquid. When you shake the jar, everything swirls around together. Then the stars slowly float to the surface, and the leaves settle to the bottom.

Rico mixed up 500 cc of salt solution. He weighed it and found its mass to be 585 g. Will his shake-up toy work the way he wants it to if he uses this salt solution? Why or why not?

Remember, 1 ml = 1 cc.

Show your math.

Object or material	Density
Stars	1.12 g/cc
Leaves	1.25 g/cc
Salt solution	

Name _____
Period _____ Date _____

LAYERING HOT AND COLD WATER

Challenge
Create a layer of red water and a layer of blue water in your vial of plain water.

Prediction
Predict and draw the order of layers in a successfully layered vial.

Conduct the investigation
1. Draw and label your successful layers.

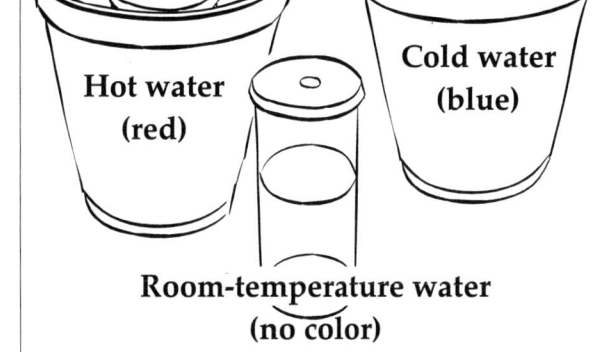

Hot water (red)   *Cold water (blue)*   *Room-temperature water (no color)*

2. Which color is densest? _____

3. Which color is least dense? _____

4. What happens after the layered vial sits for 5 minutes? Explain why.

   _____
   _____
   _____

5. What do you think would happen if you placed the layered vial in a cup of hot water 2 cm deep? Explain why.

   _____
   _____
   _____

6. Explain the relationship between temperature and density.

   _____
   _____
   _____

FOSS Weather and Water Course
© The Regents of the University of California
Can be duplicated for classroom or workshop use.

Investigation 5: Convection
Student Sheet

Name _____

Period _____ Date _____

CONVECTION CHAMBER

Part 1: Draw what you observed in the convection chamber.

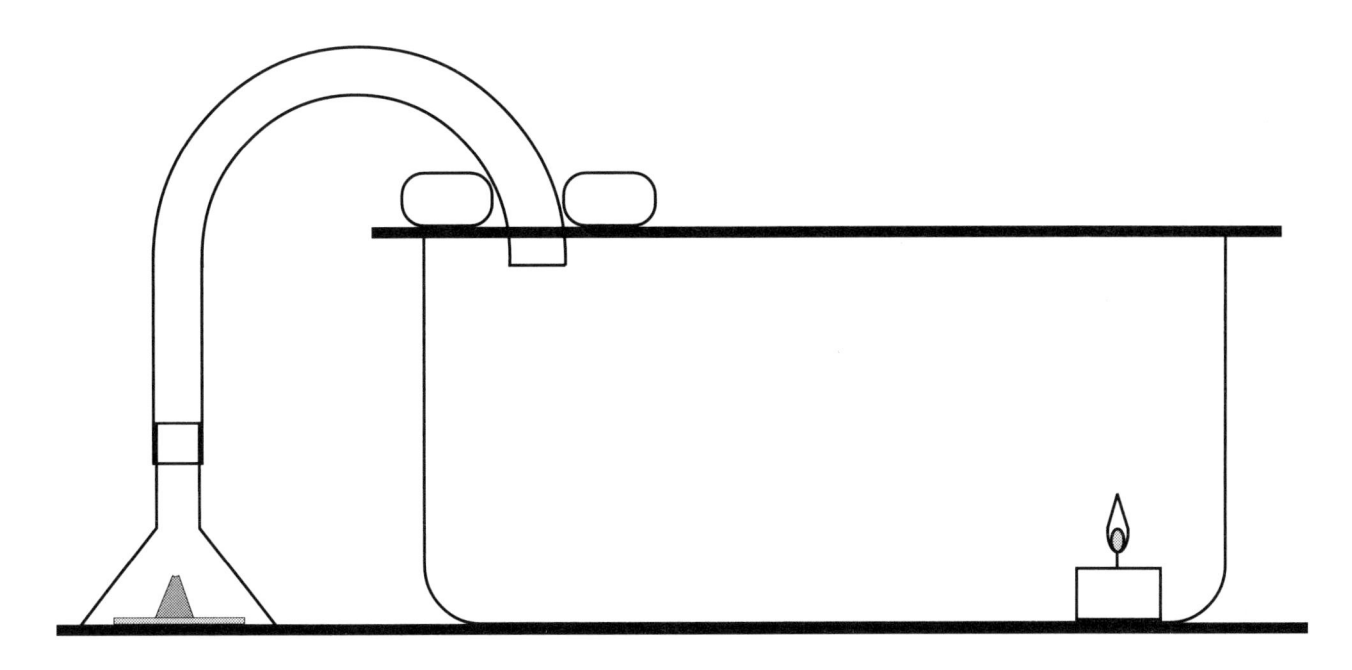

Part 2: Think about convection.

1. Explain how convection occurs in a convection chamber.

   _____
   _____
   _____
   _____
   _____

2. Explain heat transfers in a convection cell on Earth.

   _____
   _____
   _____
   _____
   _____
   _____

FOSS Weather and Water Course
© The Regents of the University of California
Can be duplicated for classroom or workshop use.

Investigation 5: Convection
Student Sheet

Name _____
Period _____ Date _____

RELATIVE HUMIDITY

Relative humidity is a comparison of the amount of water vapor in the air and the amount of water vapor needed to saturate the air at a particular temperature. Relative humidity is a percentage.

Grams of water vapor needed to saturate 1 kg of air at various temperatures

Temp.	Water vapor (g)
–40°C	0.1
–30°C	0.3
–20°C	0.8
–10°C	2.0
0°C	3.5
5°C	5.0
10°C	7.0
15°C	10.0
20°C	14.0
25°C	20.0
30°C	26.5
35°C	35.0
40°C	47.0

Example
Look at the chart on the right. If a kilogram of air at 5°C contains 5 grams (g) of water vapor, the air is saturated. The relative humidity at 5°C is 100%.

If that kilogram of air warms up to 15°C, it will need 10 g of water vapor to saturate it. But the air has only 5 g of water vapor, only half enough to saturate the air. The relative humidity at 15°C is 50%.

1. What is the relative humidity of a kilogram of air at 25°C that contains
   a. 20 g of water vapor?

   b. 5 g of water vapor?

   c. 10 g of water vapor?

   d. 16 g of water vapor?

2. A kilogram of air contains 7 g of water vapor. Its relative humidity is 50%. What is the temperature of the air? At what temperature would the air reach a relative humidity of 100%?

3. A kilogram of air has a relative humidity of 100%. It contains 2 g of water vapor. What will the relative humidity be when the air warms to 25°C?

FOSS Weather and Water Course
© The Regents of the University of California
Can be duplicated for classroom or workshop use.

Investigation 6: Water in the Air
Student Sheet

Name _____

Period _____ Date _____

RESPONSE SHEET—WATER IN THE AIR

Christine and Ingrid trotted to the sideline after a tough soccer workout. Both were dripping with sweat. Christine said to Ingrid,

I have heard that sweating helps keep you cool when you are working hard. Could that be right?

Ingrid responded,

I think it has something to do with condensation, but I'm not sure how it works.

What would you tell the girls to help them better understand sweating and cooling?

_____
_____
_____
_____
_____
_____
_____
_____
_____
_____
_____
_____
_____
_____

Name _____

Period _____ Date _____

DEW-POINT QUESTIONS

1. Dew point is monitored by meteorologists. Why do you think meteorologists are interested in dew point?

   _____

   _____

   _____

   _____

2. Under what conditions would dew *not* form?

   _____

   _____

   _____

3. What do you think would happen to water vapor that condenses on a surface that has a temperature below 0°C? What is it called?

   _____

   _____

   _____

   _____

4. People who wear glasses often see condensation on their lenses when they walk from a cold, outdoor environment into a warm house. Why does that happen? How could they prevent it?

   _____

   _____

   _____

   _____

   _____

5. Do you think dew point is always the same? How could you find out? (Write your answer on the back of this sheet.)

FOSS Weather and Water Course
© The Regents of the University of California
Can be duplicated for classroom or workshop use.

Investigation 6: Water in the Air
Student Sheet

Name _____

Period _____ Date _____

PRESSURE/TEMPERATURE DEMONSTRATION

Question

What happens to the temperature of a gas if you squeeze it into a smaller volume?

Materials

- Soda bottles
- Soda-bottle pump
- Liquid-crystal thermometers
- Masking tape

Observations and conclusion

1. Record your observations and conclusion.

   _____
   _____
   _____
   _____
   _____

2. Knowing what you do about the movement of molecules in a gas, explain the change in temperature as you changed the gas volume by squeezing the bottle or when additional air was forced into the volume of the bottle.

   _____
   _____
   _____
   _____

3. How is the bottle demonstration similar to what happened inside a syringe when you pushed in the plunger?

   _____
   _____
   _____

4. How do you think the temperature changes inside of the syringe when the air is compressed with the plunger?

   _____
   _____

FOSS Weather and Water Course
© The Regents of the University of California
Can be duplicated for classroom or workshop use.

Investigation 6: Water in the Air
Student Sheet

Name _____

Period _____ Date _____

WEATHER-BALLOON SIMULATION

View and compare the weather-balloon launches for Chicago and Phoenix. Answer these questions.

1. What was the trend in air pressure as altitude increased in Chicago? In Phoenix?

   _____

   _____

2. Describe the temperature trends in both Chicago and Phoenix. Was the trend the same in both cities?

   _____

   _____

   _____

   _____

3. Which weather factors varied the most between Chicago and Phoenix?

   _____

   _____

4. Do you think it might be a cloudy day in Chicago? What evidence do you have? At what altitude would you see clouds?

   _____

   _____

   _____

5. Do you think it might be a cloudy day in Phoenix? What evidence do you have? At what altitude would you see clouds?

   _____

   _____

   _____

FOSS Weather and Water Course
© The Regents of the University of California
Can be duplicated for classroom or workshop use.

Investigation 6: Water in the Air
Student Sheet

Name _____
Period _____ Date _____

UPPER-AIR SOUNDING GRAPH

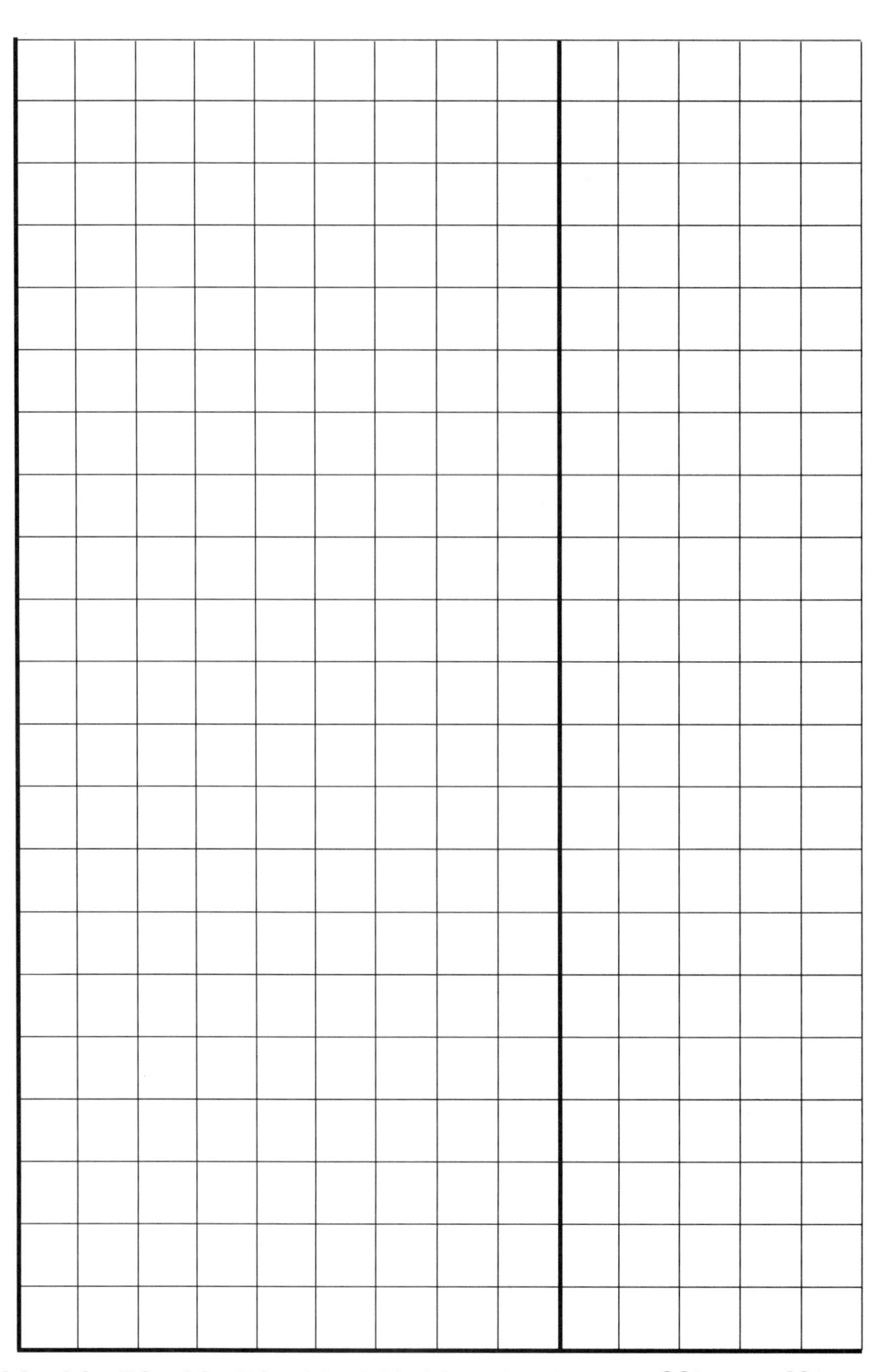

-90° -80° -70° -60° -50° -40° -30° -20° -10° 0° 10° 20° 30° 40° 50°

FOSS Weather and Water Course
© The Regents of the University of California
Can be duplicated for classroom or workshop use.

Investigation 6: Water in the Air
Student Sheet

Name _____

Period _____ Date _____

TEMPERATURE NUMBER LINE

FOSS Weather and Water Course
© The Regents of the University of California
Can be duplicated for classroom or workshop use.

Investigation 6: Water in the Air
Student Sheet

Name _____

Period _____ Date _____

WATER-CYCLE GAME

Directions: As you move through the water cycle, keep track of where you go.

	Round 1 locations	**Round 2 locations**	**Round 3 (global warming)**
Stop 1	_____	_____	_____
Stop 2	_____	_____	_____
Stop 3	_____	_____	_____
Stop 4	_____	_____	_____
Stop 5	_____	_____	_____
Stop 6	_____	_____	_____
Stop 7	_____	_____	_____
Stop 8	_____	_____	_____
Stop 9	_____	_____	_____
Stop 10	_____	_____	_____

Questions

1. Which location did you visit most often? _____
   _____

2. Which location did you visit the least? _____
   _____

3. Were there any locations you never visited? Which ones? _____
   _____

4. Write down one question you have about the water cycle after you finish the game.
   _____
   _____

FOSS Weather and Water Course
© The Regents of the University of California
Can be duplicated for classroom or workshop use.

Investigation 7: The Water Planet
Student Sheet

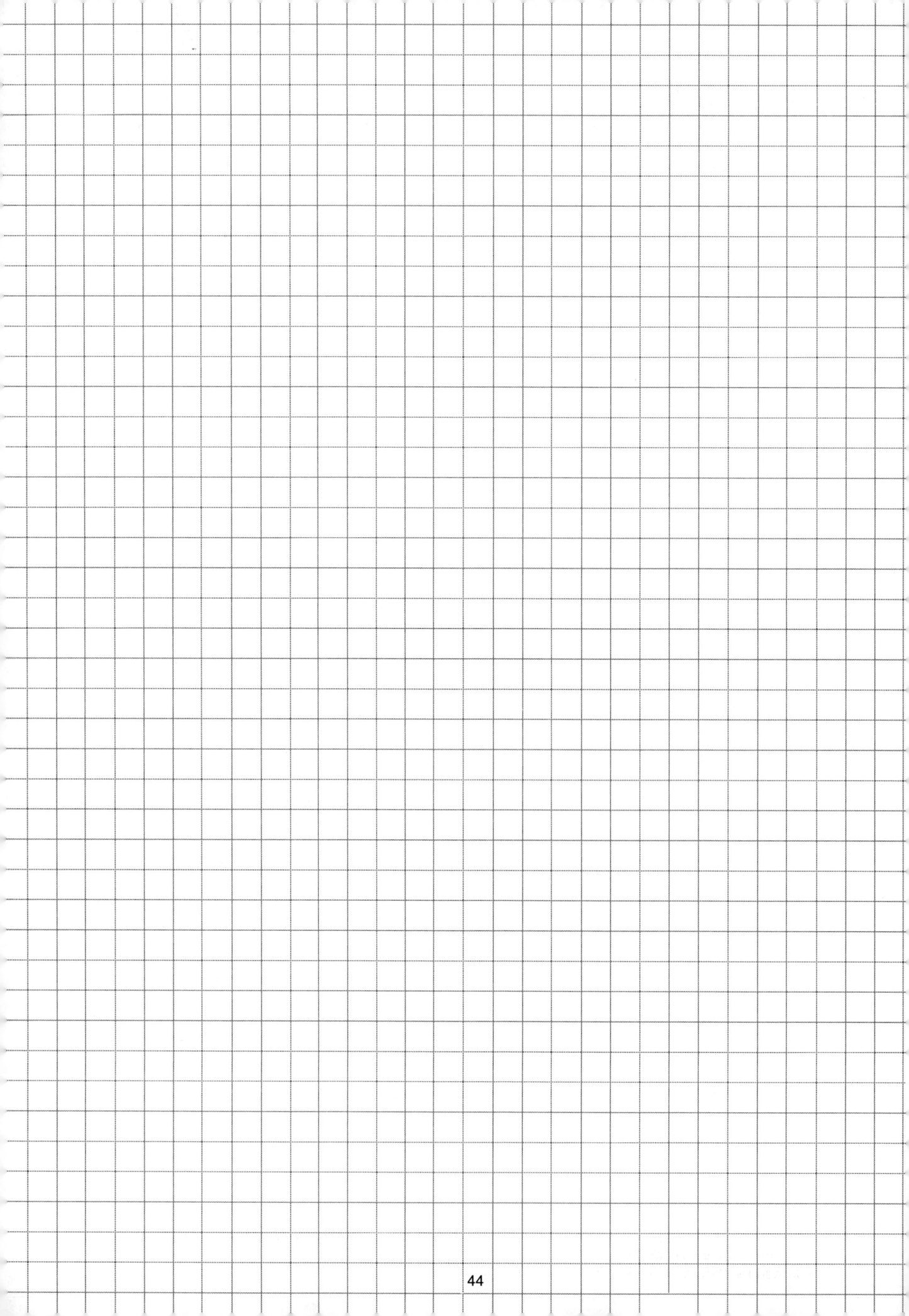

Name _____

Period _____ Date _____

PRESSURE IN A JAR

Part 1: Prediction
Predict what will happen to the water in the clear tube when the jar is squeezed.

_____
_____
_____

Part 2: Explore the jar
1. Construct a bottle-in-a-jar pressure indicator. Give it a squeeze. What happens?

_____
_____
_____

2. Why do you think it behaves that way?

_____
_____
_____
_____
_____

3. If you reduced the air pressure in the jar, what would happen to the level of water in the clear tube? Why?

_____
_____
_____
_____
_____
_____
_____

FOSS Weather and Water Course
© The Regents of the University of California
Can be duplicated for classroom or workshop use.

Investigation 8: Air Pressure and Wind
Student Sheet

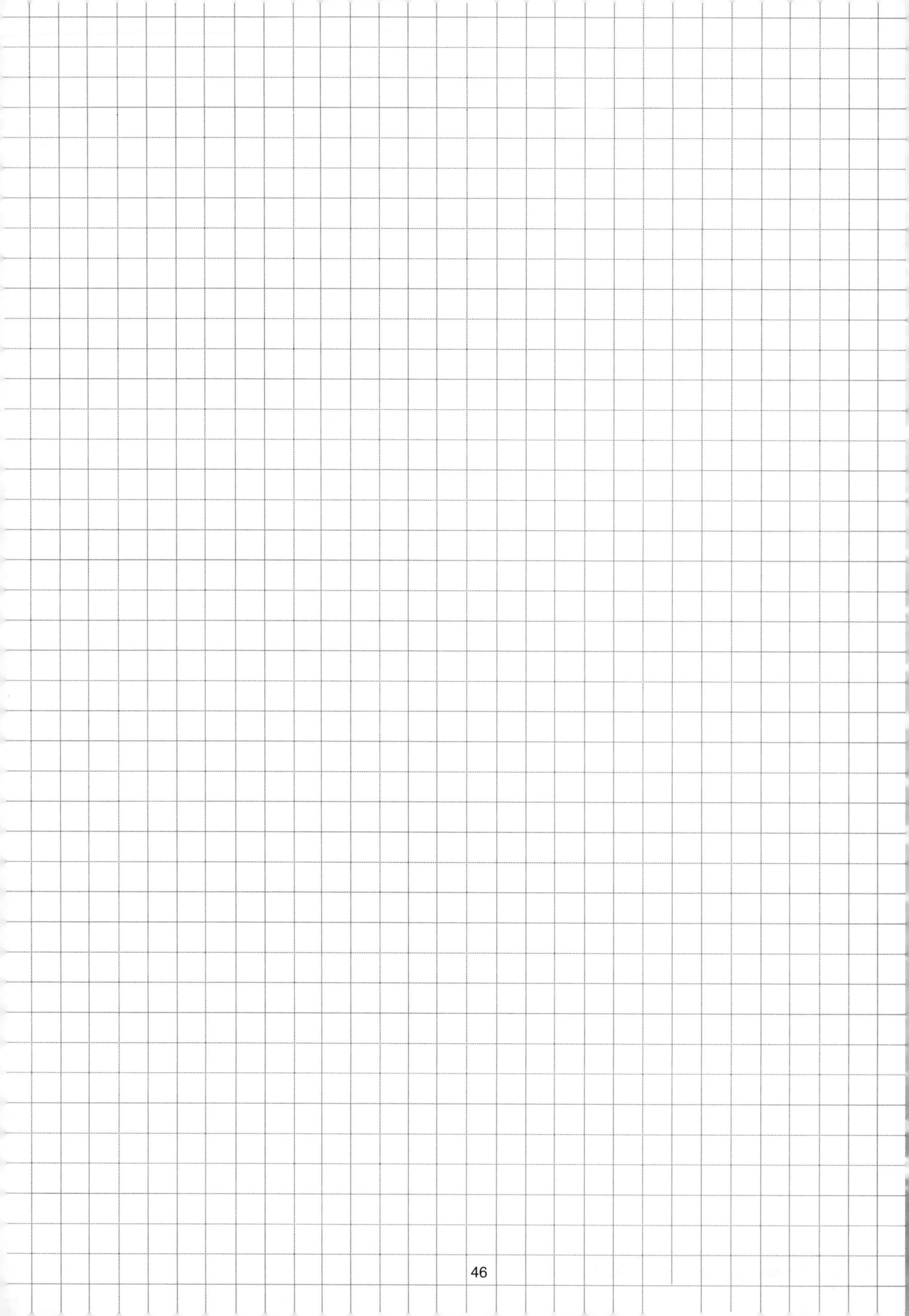

Name _____

Period _____ Date _____

RESPONSE SHEET—AIR PRESSURE AND WIND

At camp, Derek and his friends hiked up to a forest-fire lookout station at an elevation of 2445 m. On top of the mountain, Derek drank the last of the water he carried in a plastic bottle. He put the lid back on the bottle, tossed it in his day pack, and forgot about it.

When Derek returned home to Seattle 3 days later, and unpacked his stuff, he found his water bottle. It looked like this.

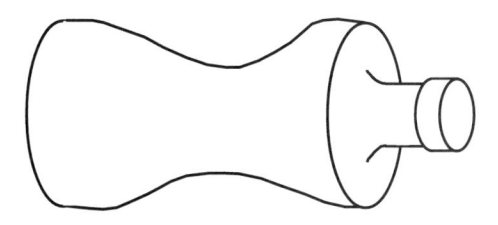

Derek thought,

I must have sat on my water bottle, or something, to squash it like that.

When he unscrewed the lid, he heard a hissing sound, and the bottle slowly returned to its proper shape. What do you think happened to Derek's bottle? Can you explain Derek's squashed bottle and the hissing sound?

_____
_____
_____
_____
_____
_____
_____
_____
_____
_____
_____
_____

FOSS Weather and Water Course
© The Regents of the University of California
Can be duplicated for classroom or workshop use.

Investigation 8: Air Pressure and Wind
Student Sheet

Name _____

Period _____ Date _____

LOCAL WINDS

View the local-winds animations on the computer and follow the directions below.

Sea Breeze

Draw the land, water, and Sun.
Show circulation by convection.
Label the low- and high-pressure areas.
Draw an arrow showing the wind direction.

Land Breeze

Draw the land, water, and Sun.
Show circulation by convection.
Label the low- and high-pressure areas.
Draw an arrow showing the wind direction.

FOSS Weather and Water Course
© The Regents of the University of California
Can be duplicated for classroom or workshop use.

Investigation 8: Air Pressure and Wind
Student Sheet

Name _____

Period _____ Date _____

Valley Breeze

Draw the mountain slope and Sun.
Show circulation by convection.
Label the low- and high-pressure areas.
Draw an arrow showing the wind direction.

Mountain Breeze

Draw the mountain slope and Sun.
Show circulation by convection.
Label the low- and high-pressure areas.
Draw an arrow showing the wind direction.

FOSS Weather and Water Course
© The Regents of the University of California
Can be duplicated for classroom or workshop use.

Investigation 8: Air Pressure and Wind
Student Sheet

MAKING AN ANEMOMETER

Materials
- 2 Index cards, 7.5 cm × 13 cm
- 2 Straws, super jumbo
- 1 Straw, jumbo
- 7 Paper clips, regular
- • Transparent or masking tape
- 1 Hole punch
- 1 Scissors
- 1 Metric ruler
- 1 Pencil

Preparation and assembly

1. Fold both edges of card 1 at a right angle. This will be easier if you draw lines 1.2 cm from the edges, pressing down hard with a pencil or ballpoint pen to crease the card.

2. Punch two holes near the top of each folded edge.

3. Lay card 2 on the template on this page. Mark the edge of the card where the ends of the lines extend beyond the card. Draw lines between the marks and write the numbers on the lines.

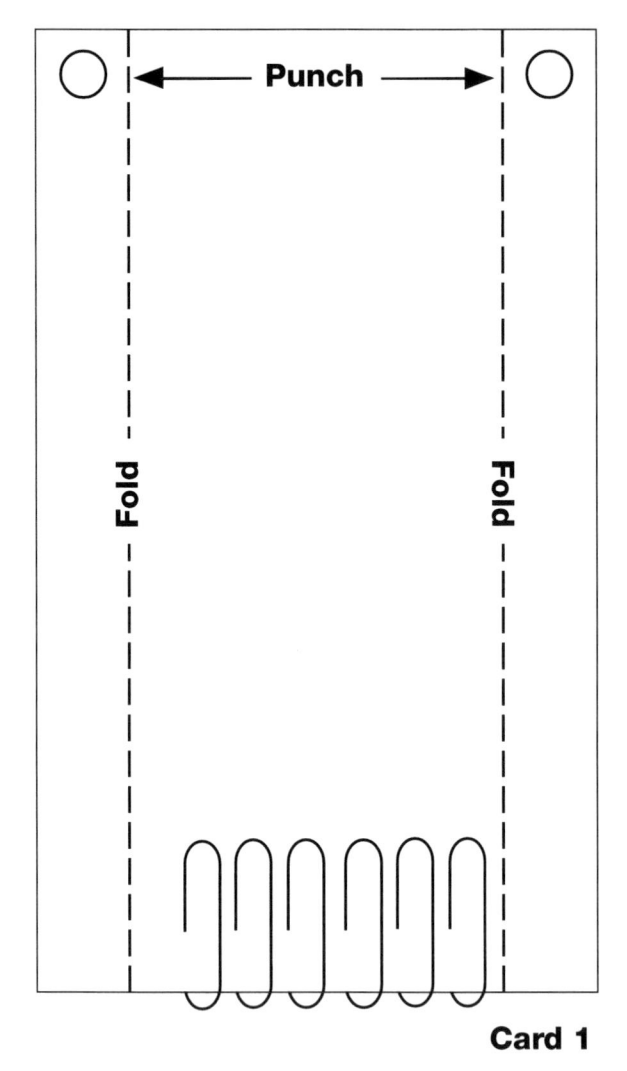

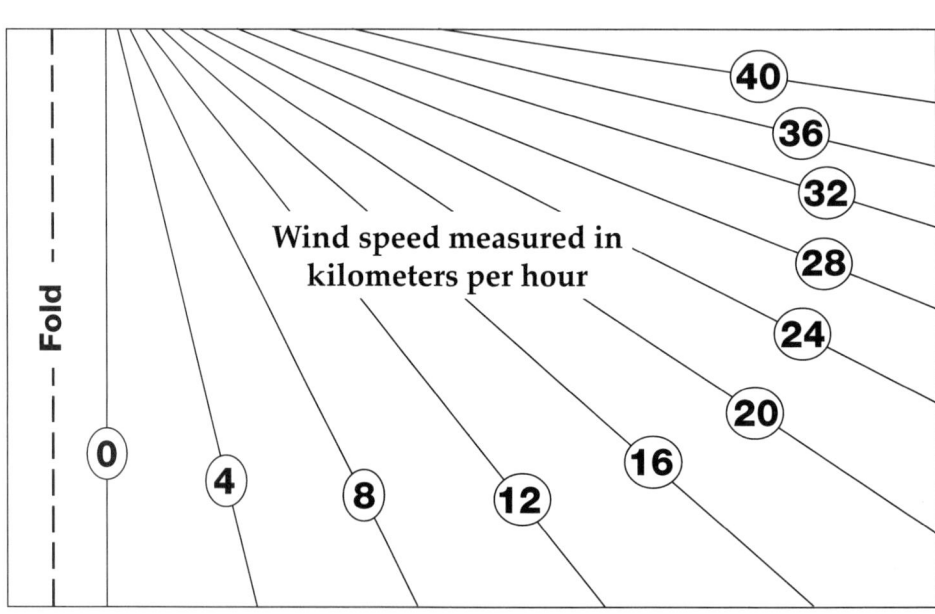

4. Fold the edge of the card at a right angle.

Card 1

Punch / Fold / Fold

Card 2

Wind speed measured in kilometers per hour

0, 4, 8, 12, 16, 20, 24, 28, 32, 36, 40

FOSS Weather and Water Course
© The Regents of the University of California
Can be duplicated for classroom or workshop use.

Investigation 8: Air Pressure and Wind
Student Sheet

5. Punch a hole in two super jumbo straws about 2.5 cm from one end.

6. Cut a clear jumbo straw to a length of 12 cm. Slide the clear straw through the holes in the super jumbo straws. Tape the two super jumbo straws together top and bottom.

7. Slide card 2 between the two super jumbo straws. Push it up as far as it will go toward the jumbo straw. Tape the folded-over edge to hold card 2 in place.

8. Slide card 1 onto the jumbo straw. Attach six paper clips to the bottom of card 1, and one on the end of the jumbo straw to keep card 1 from sliding off the straw.

9. Hold the anemometer in the wind so that the wind hits the broad side of card 1. Read the wind speed in kilometers per hour from card 2.

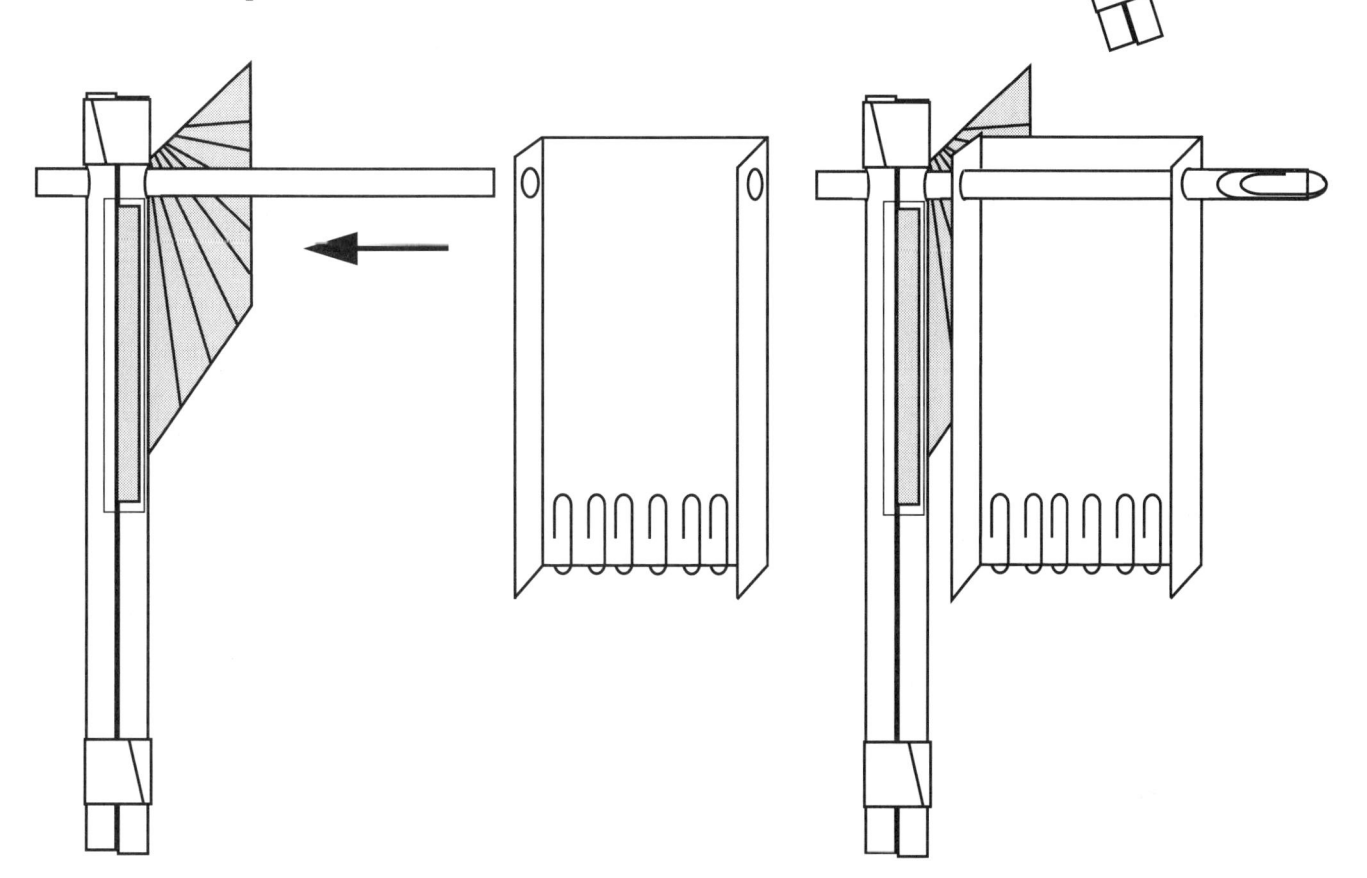

FOSS Weather and Water Course
© The Regents of the University of California
Can be duplicated for classroom or workshop use.

Investigation 8: Air Pressure and Wind
Student Sheet

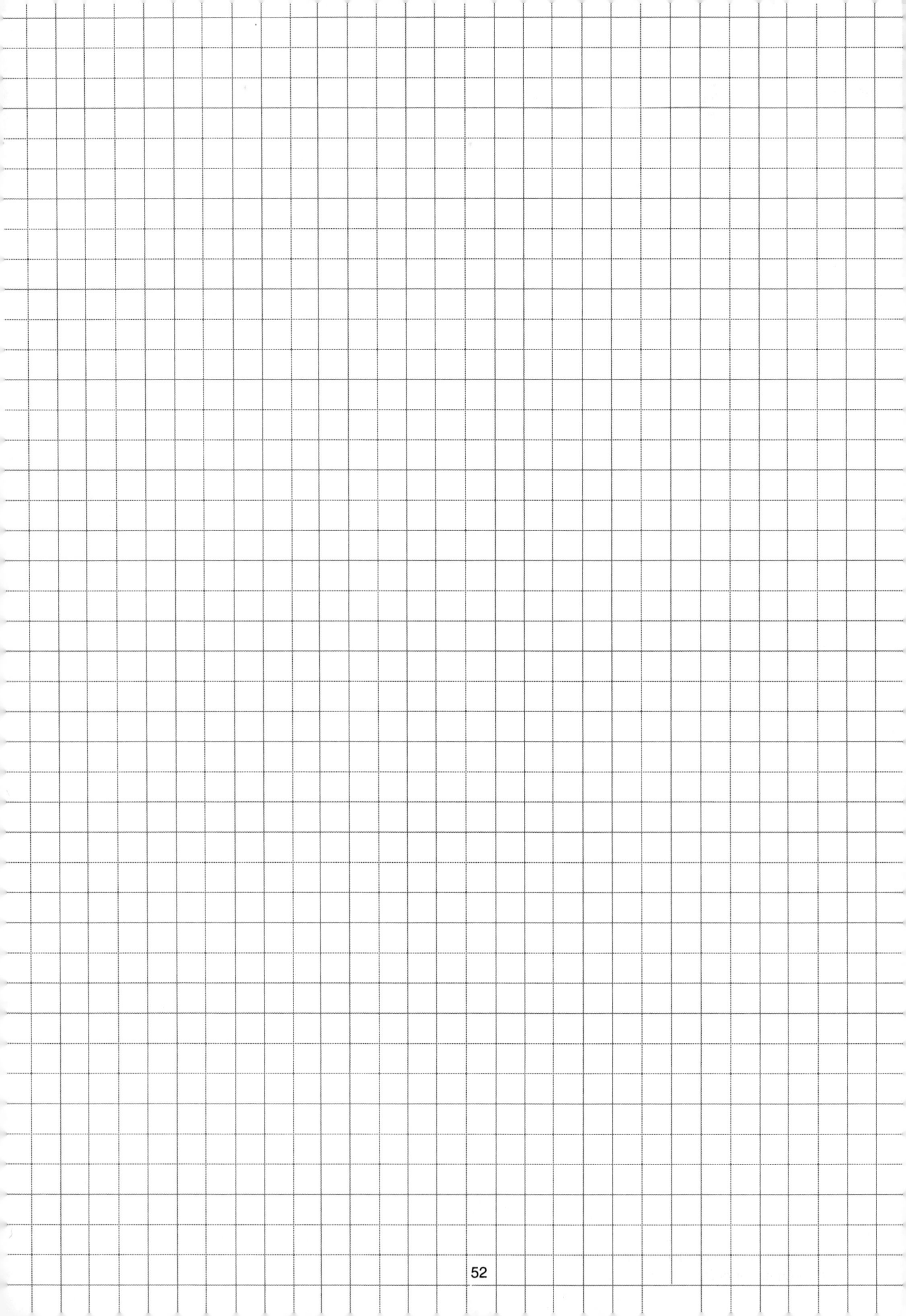

PRESSURE MAP OF THE U.S.

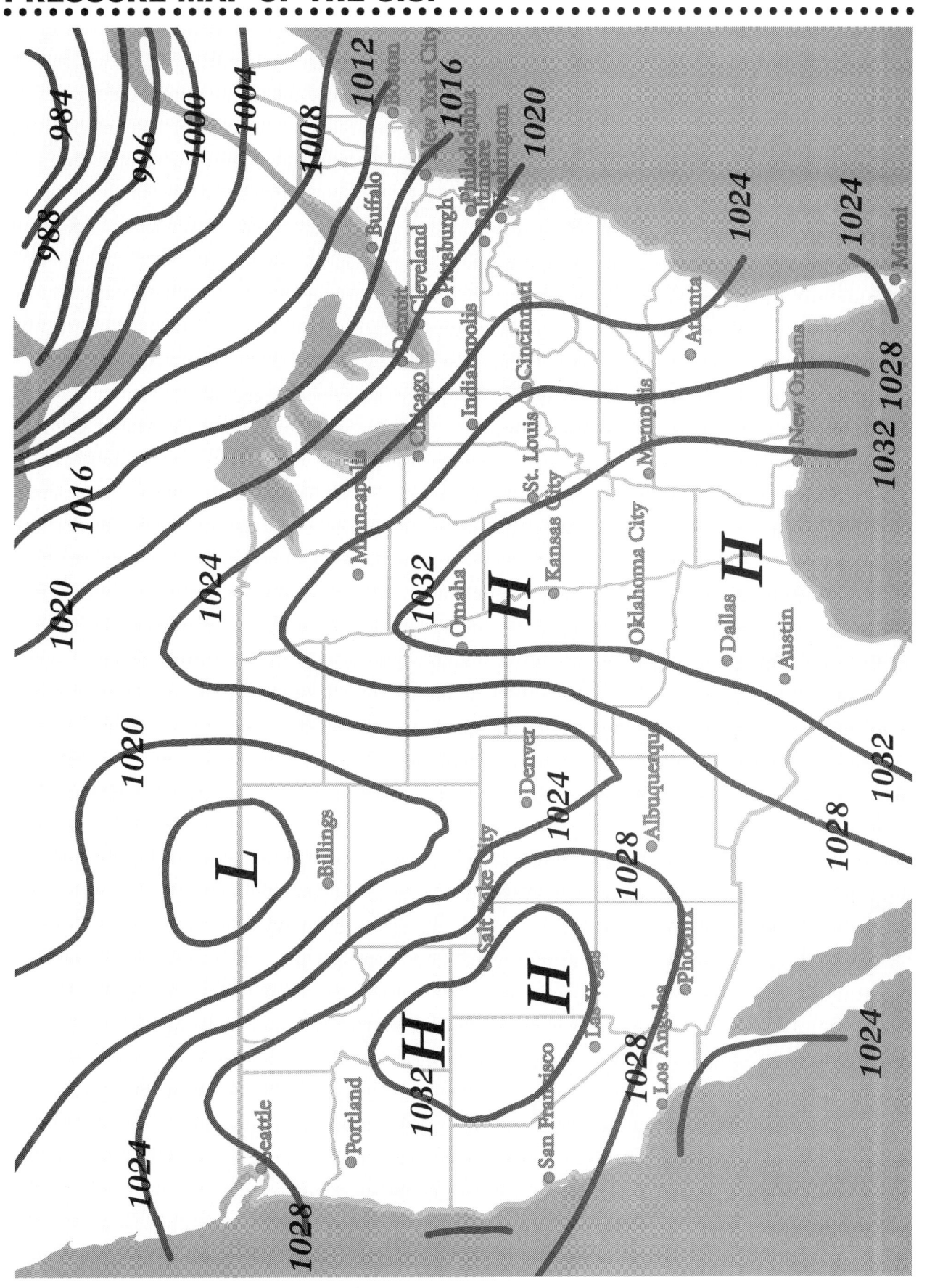

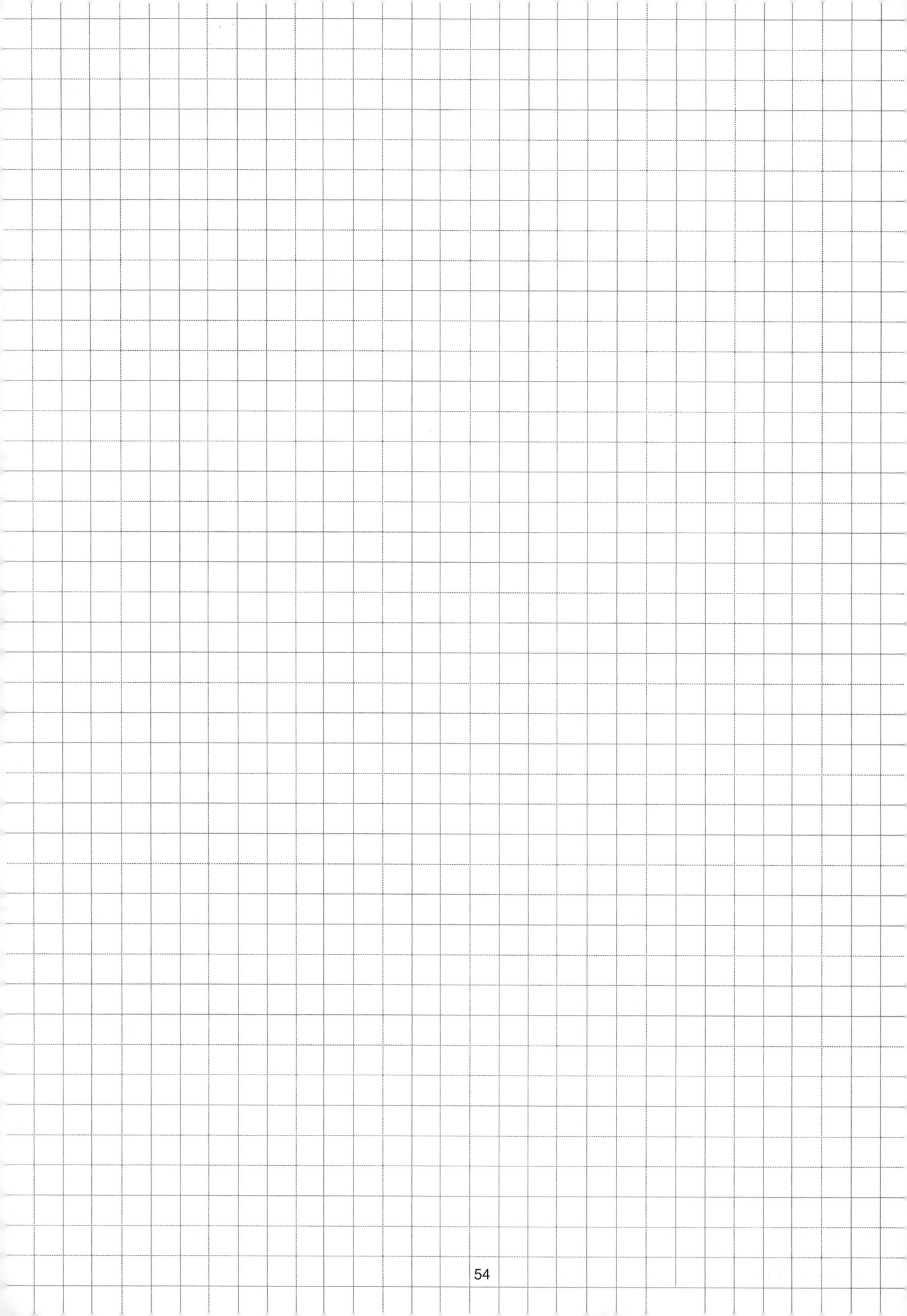

Name _____

Period _____ Date _____

SOLAR-BALLOON OBSERVATIONS

1. Describe the bag's properties before it is used to model an air mass.

   _____
   _____
   _____

2. Describe what happened to the bag while it was outside.

   _____
   _____
   _____
   _____
   _____
   _____

3. How is the black-bag model like a real air mass and how is it different?

   _____
   _____
   _____
   _____
   _____
   _____

4. What questions do you have about the formation of air masses after observing this model?

   _____
   _____
   _____
   _____

FOSS Weather and Water Course
© The Regents of the University of California
Can be duplicated for classroom or workshop use.

Investigation 9: Weather and Climate
Student Sheet

Name _____

Period _____ Date _____

READING WEATHER MAPS

(pg. 87) *(pg. 88)*

Use the *Sample Weather-Map Symbol* sheet and the *Surface Observations* map in the resources book to complete this sheet.

1. Find the weather-station data for San Francisco, California. Use the weather-map symbol to figure out the following:

   Temperature (°F) _____   Air pressure (mb) _____

   Wind direction _____   Cloud cover _____

   Wind speed (knots) _____

2. Which cities are experiencing haze? _____
   _____

3. Which cities are experiencing rain? _____
   _____

4. Which cities are experiencing fog? _____
   _____

5. Which city has the highest air pressure? How high is it? The lowest? How low is it?
   _____
   _____

6. Which areas of the United States are experiencing the warmest temperatures?
   _____
   _____

7. List two other observations about the weather across the United States for this date.
   _____
   _____

FOSS Weather and Water Course
© The Regents of the University of California
Can be duplicated for classroom or workshop use.

Investigation 9: Weather and Climate
Student Sheet

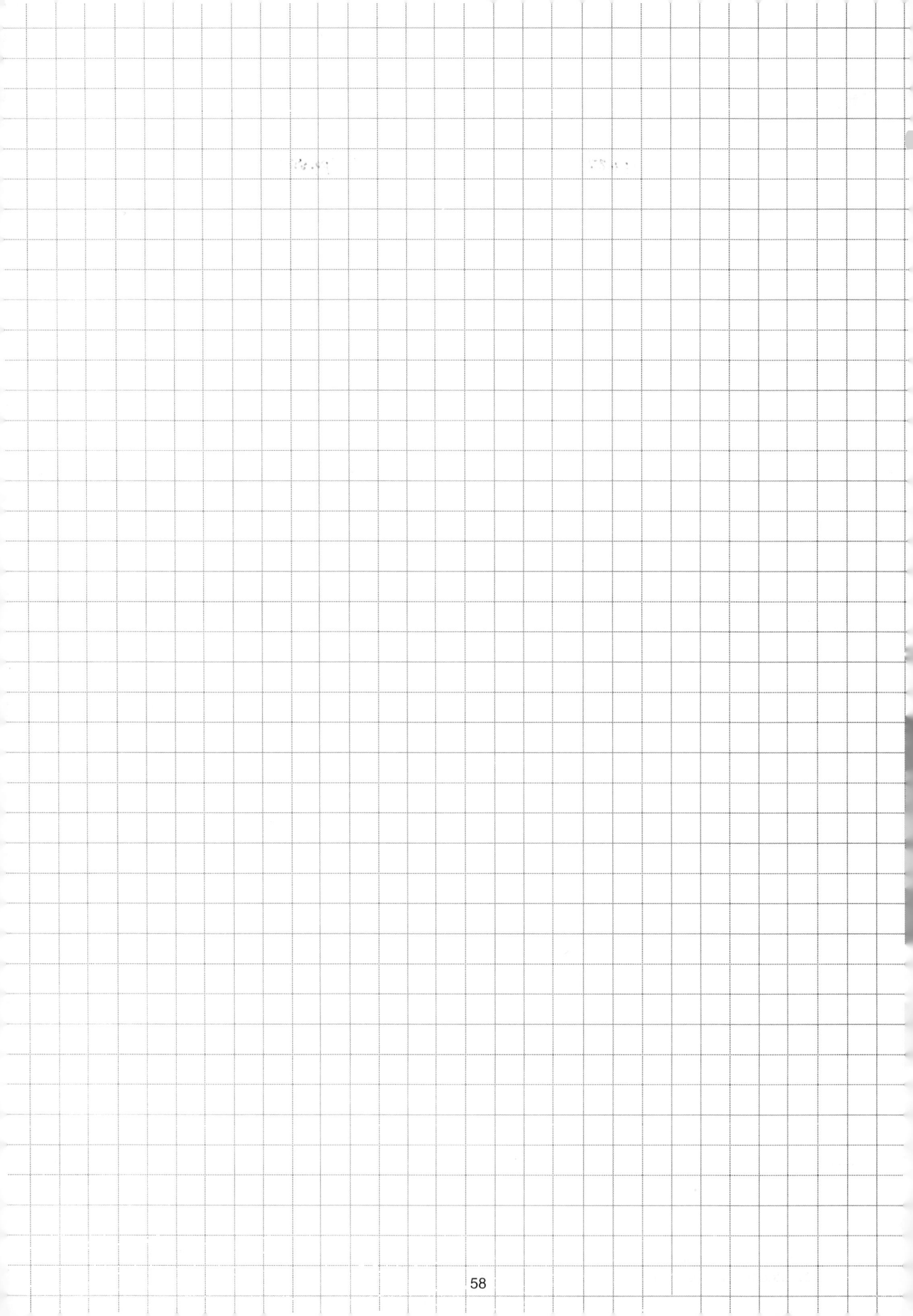

Name _____
Period _____ Date _____

RESPONSE SHEET—WEATHER AND CLIMATE

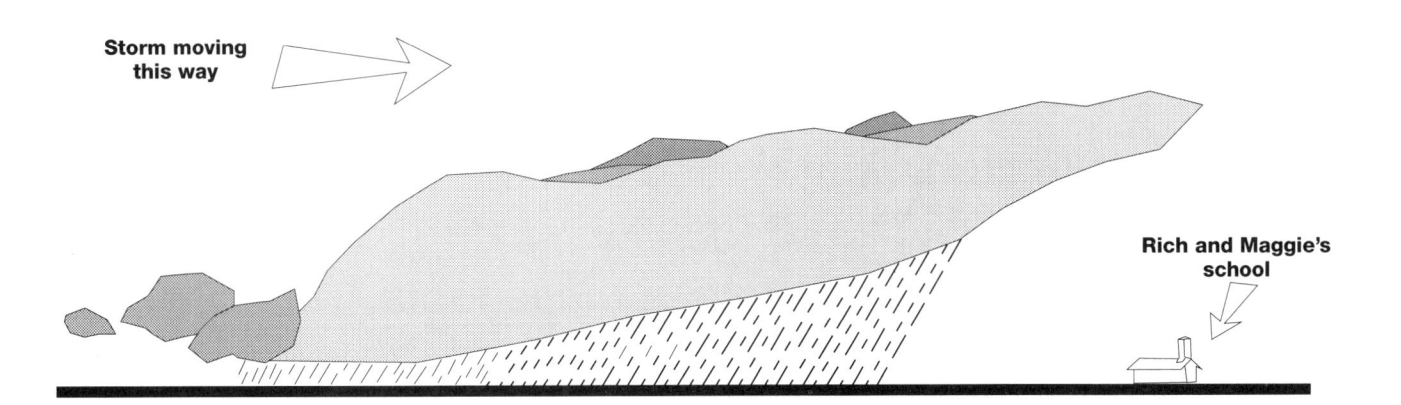

[Diagram: Storm moving this way → with clouds, rain, and Rich and Maggie's school]

Rich and Maggie saw on the news that a rainstorm was heading their way.

"It is really cold today," Rich said. "That must be a cold front coming our way."

"I'm not sure," said Maggie. "Here's a picture of what I think the storm looks like. I'm trying to figure out where the warm air mass is and where the cold air mass is."

Can you help Rich and Maggie understand what is going on with the weather to create the big, wet storm?

_____
_____
_____
_____
_____
_____
_____
_____
_____
_____
_____
_____
_____

FOSS Weather and Water Course
© The Regents of the University of California
Can be duplicated for classroom or workshop use.

Investigation 9: Weather and Climate
Student Sheet

ASSESSMENT—GENERAL RUBRIC

4 The student uses two or more facts to explain a bigger idea by making connections between those facts. All of the information is correct, and the connections and conclusions are correct.

3 The student uses two or more facts to attempt to explain a bigger idea by making connections between those facts. The facts or the connections have minor errors.

2 The student provides two or more facts that are related to the task or questions asked, but does not make any connections between the facts.

1 The student provides one fact that is related to the task or question asked.

0 The student does not answer the question, does not complete the task, or gives an answer that has nothing to do with what was asked.